SpringerBriefs in Electrical and Computer Engineering

SpringerBriefs in Computational Electromagnetics

Series editors

K.J. Vinoy, Bangalore, India
Rakesh Mohan Jha (Late), Bangalore, India

More information about this series at http://www.springer.com/series/13885

Satyakesh Dubey · Naina Narang
Parmendra Singh Negi · Vijay Narain Ojha

LabVIEW Based Automation Guide for Microwave Measurements

 Springer

Satyakesh Dubey
Microwave Standards, TFEEMD
CSIR-National Physical Laboratory of India
New Delhi
India

Naina Narang
CSIR-National Physical Laboratory of India
New Delhi
India

Parmendra Singh Negi
CSIR-National Physical Laboratory of India
New Delhi
India

Vijay Narain Ojha
Microwave Standards, TFEEMD
CSIR-National Physical Laboratory of India
New Delhi
India

ISSN 2191-8112   ISSN 2191-8120   (electronic)
SpringerBriefs in Electrical and Computer Engineering
ISSN 2365-6239   ISSN 2365-6247   (electronic)
SpringerBriefs in Computational Electromagnetics
ISBN 978-981-10-6279-7   ISBN 978-981-10-6280-3   (eBook)
https://doi.org/10.1007/978-981-10-6280-3

Library of Congress Control Number: 2017950254

Printed on acid-free paper

This Springer imprint is published by Springer Nature
The registered company is Springer Nature Singapore Pte Ltd.
The registered company address is: 152 Beach Road, #21-01/04 Gateway East, Singapore 189721, Singapore

Preface

At times, the proper use of a synchronized set of instruments becomes indispensable for scientists and engineers in developing an accurate measurement or test setup. Also, the need for automated execution is always felt in day-to-day work in measurement laboratories for faster execution and reduced manual interference in the measurement process. If we consider the microwave measurements, in particular, it involves the use of advanced electronics, embedded systems, and microcontrollers. Proper use of such advanced devices in synchronization becomes important for the development of measurement facilities in laboratories such as NMIs and defense research institutes, where accuracy and precision play one of the most crucial roles. This book guides the readers for developing a standalone graphical user interface (GUI) and explains the software development process for effective microwave measurements. An effort is made to develop such a software application, which is independent of the measurement setup and measurand, i.e., the quantities intended to be measured. The critical considerations for quality assurance of the developed system are studied and illustrated in a way, which is suitable for beginners and experts in measurement laboratories. A worked out example is used throughout the book, which is used for various test procedures with different sources, receivers, and devices under test (DUT). The measurement results are illustrated for comparison of automated and manual execution of the measurement process. The approaches required for optimizing the tests for time, reliability, and throughput are formulated and used during the development process. All in all, the book is an easy guide for scientists and engineers for development of automated laboratory measurements with quality assurance, efficiency optimization, and adherence to the quality standards for using the software in metrological applications.

New Delhi, India

Satyakesh Dubey
Naina Narang
Parmendra Singh Negi
Vijay Narain Ojha

Contents

List of Figures

List of Tables

About the Authors

Dr. Satya Kesh Dubey dubeysk@nplindia.org, was born in 1984 in Madhuban, Mau (U.P.), India. He completed his Ph.D. in Electronics from Allahabad University, Allahabad, India, with eight international journal papers. He joined National Institute of Technology Raipur (C.G.) as lecturer in 2009. He served Institute for Plasma Research as postdoctoral fellow for a year during 2009–2010 in microwave diagnostic group. Again he joined National Aerospace Laboratories, Bangalore as scientist fellow in computational electromagnetic group founded by the eminent scientist late Dr. R.M. Jha. He joined National Physical Laboratory, New Delhi as Scientist in Microwave Activity. He has guided more than 15 M. Tech. students with four enrolled Ph.D. scholars. He has published 35 papers in different international journals and conference proceedings with h- index 8 and i- index 6. His current area of research is biological effect of EM radiations, electreomagnetic-induced transparency, E-field probes and sensors, SAR probes, microstrip antenna, millimeter wave, and computational modeling of biological tissue in EM radiations.

Naina Narang nainanarang@nplindia.org, is a Computer Science Graduate from Kurukshetra University, India and is working as Assistant Professor at Department of Computer and Communication Engineering, School of Computing and Information Technology Manipal University, Jaipur, India. She had worked with Dr. Satya Kesh Dubey as Ph.D. research scholar for the establishment of Standards of Specific Absorption Rate (SAR) at National Physical Laboratory. She has a keen interest in instrument control, LabVIEW programming for automation, computational and numerical techniques for electromagnetics.

Parmendra Singh Negi psnegi@nplindia.org, Head Microwave Standards, TFEEMD, CSIR-National Physical Laboratory, New Delhi. He has an experience of more than 30 years and expertise in attenuation and impedance measurement. He has piloted/participated in several key/supplementary comparisons at international level. He had developed mismatch standards, WBCO-based primary standard for attenuation with high accuracy and is author of more than 100 journal and conference papers.

Dr. Vijay Narain Ojha vnojha@nplnida.org, Head, TFEEM Division. He worked as Scientist at CSIR-National Physical Laboratory, New Delhi for more than 30 years and now leads the Electrical and Electronics Standards in India. He has keen research interest in low-temperature microwave measurements, Josephson junction, quantum Hall effect and Watt balance, and is author of more than 100 journal and conference papers.

Chapter 1
Introduction

In a very conventional approach, it can be stated that every microwave measurement consists of a transmitter, a receiver and a device under test (DUT). The transmitter can be an economical signal generator or a stable and precise radio frequency (RF) source. Similarly, the receiver can be a spectrum analyzer, vector network analyzer (VNA), power meter with RF power sensor, or a receiver specific for particular measurement such as TEGAM VM-7 attenuation measurement system. The DUT, on the other hand, is the instrument or device whose electrical properties are intended to be measured. In a microwave metrology laboratory, a DUT can be a fixed attenuator, step attenuator, airline, waveguide, or any other component such as mixer, directional coupler, power divider or an amplifier. For any microwave measurement process, these instruments can be configured for automatic operation using the various commercially available application development platforms such as MATLAB, Agilent VEE Pro, or National Instruments (NI) LabVIEW. The automated software solution for a measurement process may minimize the human involvement and thus reducing the errors and improving efficiency.

In the automated measurement setup, the instruments are given a remote control; therefore, an appropriate step by step approach is to be used for software development for proper execution of the measurement process. The main considerations for a developer that arise before the development process of automation software are as follows:

- *What* is to be measured and *what* are the instruments to be controlled?
- *How* the control flows within the experimental setup?
- *What* development techniques are to be followed for producing quality results?
- *Does* the approach used benefit the end user as compared to a manual execution?

This book addresses these questions in detail with worked out examples of some important microwave measurements for direct reference. The intended readers are those who are responsible for regular measurements and calibrations mainly in the

© The Author(s) 2018
S. Dubey et al., *LabVIEW Based Automation Guide for Microwave Measurements*,
SpringerBriefs in Computational Electromagnetics,
https://doi.org/10.1007/978-981-10-6280-3_1

metrology laboratories and wish for a faster execution of the experiment with the use of customized automation software.

This book is divided into six chapters. Chapter 2 provides the basic knowledge about the LabVIEW and its programming for instrument control. LabVIEW, i.e., Laboratory Virtual Instrument Engineering Workbench, is a graphical programming language developed by NI for customized solution in test and measurement laboratories. The chapter explains the control of the commonly used instruments and devices in microwave measurements using LabVIEW. It will be seen that LabVIEW provides an easy and fast way for automating measurements. It can be used to control the instruments using bus level communications in a more user-friendly way as compared to C, .NET, or MATLAB codes.

Chapter 3 introduces the readers with the common microwave measurements, their execution, uncertainty evaluation, and applications.

Chapter 4 describes the detailed procedure of the software development process and addresses the major considerations for ensuring the software validity and compliance with the quality standards that are required to use the software in metrological applications. The book guides the engineers and technicians for implementing complete calibration software with enough details to start from scratch.

Chapter 5 discusses the measured results and evaluation of uncertainty in the measurement. Comparison of the results with manual execution is reported for every example to ensure the benefit of the end user. Data collection, analysis, and report generation are explained in detail.

Chapter 6 concludes carried out study. A completely automated calibration process for microwave measurements is created and concluded.

An ample number of examples, illustrations, flowcharts, measurement results, and screen shots of a worked out automation software for microwave measurement are incorporated to provide real-life experience to the readers.

Chapter 2
LabVIEW Programming

This chapter provides the basic knowledge about the LabVIEW and its programming for instrument control. LabVIEW, i.e., Laboratory Virtual Instrument Engineering Workbench, is a graphical programming language used by test and measurement researchers and engineers for building customized software solution to their experiments [1–3]. It has an easy approach and can be used with ease. The chapter provides sufficient information to the reader for starting basic programming in LabVIEW for instrument control. This chapter can be used as a guide to connect the instrument to the computer system and basic input/output instructions. The programming in LabVIEW environment is based *data-flow* method and is mostly graphical and easy to understand.

The chapter will provide the brief introduction to LabVIEW, LabVIEW programming and its instrument control capabilities. It will cover the basic features to give an understanding of the LabVIEW programming language and its environment. Examples for controlling the RF instruments explained in this chapter will help to develop calibration suite for attenuation given in Chap. 4.

2.1 Basic Concepts

In this section, we will discuss the basic concepts about LabVIEW and its programming technique. The section is self-contained and is sufficient to understand the basics of instrument control using LabVIEW.

2.1.1 Virtual Instruments (VI)

Virtual Instrument (VI) is the file on which LabVIEW coding is done. It comprises two components—front panel and block diagram. The front panel is used to provide

© The Author(s) 2018
S. Dubey et al., *LabVIEW Based Automation Guide for Microwave Measurements*,
SpringerBriefs in Computational Electromagnetics,
https://doi.org/10.1007/978-981-10-6280-3_2

the control components such as a regulatory knob, string input, numerical input or string display. The front panel serves as the graphical user interface and can be easily designed. The block diagram, on the other hand, consists of the programming functions such as algebraic operations, instrument input/output, signal processing, data communication, and error handling. The programming logic is implemented on the block diagram using the different functions available with it.

2.1.2 Front Panel

The front panel which also serves as the graphical user interface for LabVIEW application is used to simulate the input and output mechanism. The input from the user which is to be sent to the instruments can be entered through the numeric controls, text controls, or Boolean input. Similarly, for the displaying the output numeric, text, and Boolean indicators are used on the front panel. Figure 2.1 shows the numeric and string controls and indicators that can be used to design the front panel. The controls are driven by the user, and the indicators are used to display the execution results. In the figure, Numeric 1 and Numeric 2 are the numeric controls, String is the text control, Result is the numeric indicator, and String Length is the text indicator.

Fig. 2.1 Front panel

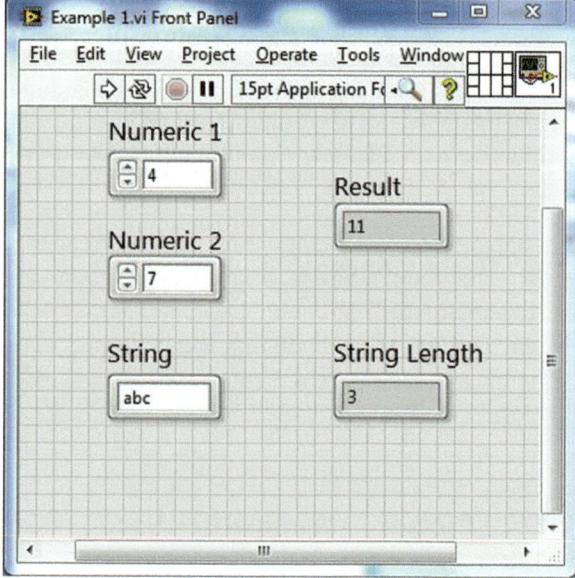

2.1.3 Block Diagram

The front panel shown in Fig. 2.1 will automatically be accompanied by the block diagram. Pressing Ctrl + E opens the block diagram. For the previous example, the node used for the controls and indicators will automatically appear on the block diagram. Figure 2.2 shows the corresponding nodes of the control and indicators used in Fig. 2.1. There are various functions available on the block diagram, for example, the arithmetic operation of addition is used to add Numeric 1 and Numeric 2. Also, programming operations on text can also be done, for example, string length function is used on String Control to calculate the string length indicated on the String Length indicator.

It can be realized by the LabVIEW user that LabVIEW provides an easy and shortcut methods to basic programming problems. As in the given example, the string length function is an easy way that takes off the burden from the programmer, unlike the other programming languages, to write a detailed code for retrieving the length of the string.

Fig. 2.2 Block diagram

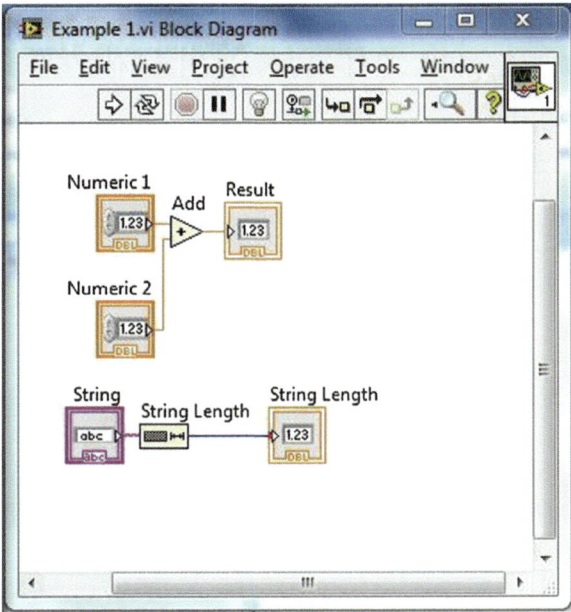

2.2 LabVIEW Features

From the introduction above, it is evident that LabVIEW is an easy-to-learn programming language based on data flow used for the implementation of the modern
measurement setup. There are some applications in which LabVIEW is the most
preferred choice, such as

- Instrument control
- Acquiring and analyzing measurement data
- Designing of embedded systems and field-programmable gate array (FPGA)
- Automated test and validation systems.

In this book, only instrument control and automated test systems are implemented and explained. You will see in Chap. 4, how an example of attenuation
measurement system is automated to provide reduced test time and efficient result
analysis.

2.3 Virtual Instrumentation

2.3.1 Virtual Instrumentation Foundation

Virtual instrumentation is a modern technique used to emulate the manual execution of a measurement with the help of a computer system [4]. The basic idea of the
virtual instrument is shown in Fig. 2.3. The virtual instrument aims to provide
better accuracy and precision with minimum manual interference in the measurement. LabVIEW is one the powerful languages which can be used to build virtual
instruments.

2.3.2 General Purpose Interface Bus (GPIB)

The General Purpose Interface Bus (GPIB) is a parallel communication interface
generally used for instrument control. This bus when connected to the PC at one

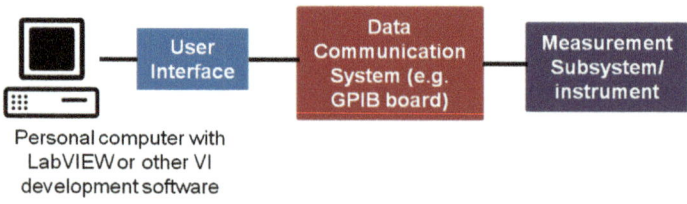

Fig. 2.3 Basic idea of virtual instrument (VI)

end and GPIB compatible instrument(s) at the other end, then the PC can be used as a system controller for the connected instrument(s). The main features of this bus are the following:

- Standardized as IEEE 488
- 8-bit parallel communication using asynchronous handshaking protocol
- The bus has 24-pin configuration where eight are data lines, five are bus management lines, three are handshake lines, and eight are ground lines
- One system controller and 14 instruments can be connected to a single GPIB
- The GPIB instruments on the bus are uniquely identified by 5-bit GPIB (Primary) address ranging between 0 and 30.

Using the NI Measurement and Automation Explorer (MAX), one can detect and configure the connected GPIB instruments on the PC. Figure 2.4 shows a GPIB instrument detected by the NI MAX. This is possibly the first step toward the learning of instrument control and virtual instrumentation.

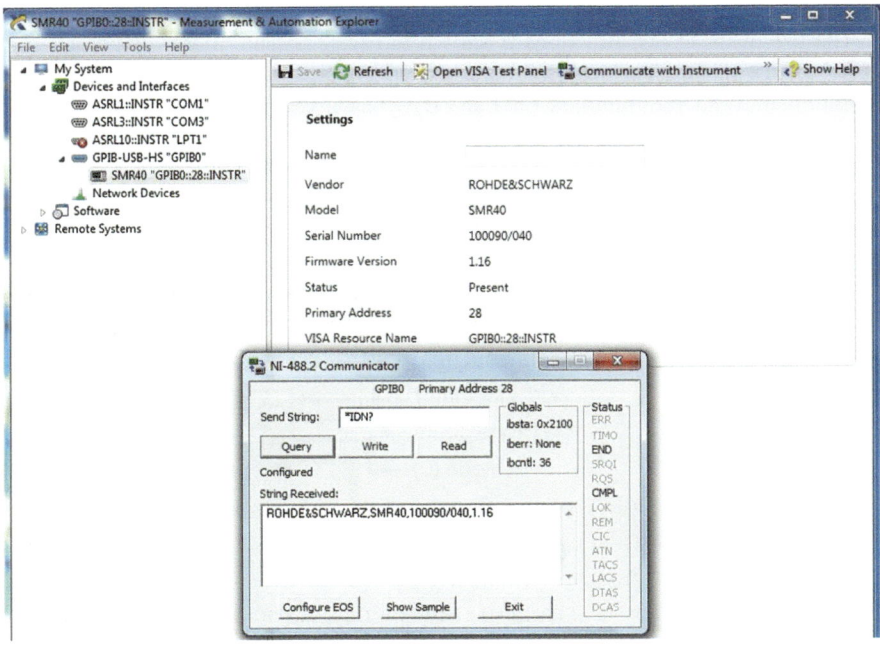

Fig. 2.4 NI measurement and automation explorer (MAX) for detecting GPIB instrument

2.3.3 Standard Commands for Programmable Instruments (SCPI)

Fig. 2.4, the GPIB instrument with primary address 28 is connected to the PC. Using the NI-488.2 Communicator, one can communicate with the instrument by sending the query. For example, when string *IDN? is sent as a query, the detailed identity of the connected instrument is returned. Here, *IDN? is an SCPI command.

2.3.4 Instrument Control in LabVIEW

To start instrument control using LabVIEW, one must be acquainted with the GPIB or Virtual Instrument Software Architecture (VISA) functions. Figure 2.5 shows the input and output wires of GPIB read and write functions. GPIB read function is used to retrieve the information from the GPIB device to the controller, i.e., the PC and GPIB write is used for sending the command from the controller to the connected GPIB instrument.

The use of these two functions is shown in Fig. 2.6. The *IDN? command sent to the source of GPIB address 28 using NI MAX is shown in Fig. 2.4. The same operation may be programmed on LabVIEW, as given in Fig. 2.6, using simple GPIB write and read functions. In the same manner, other SCPI commands can be sent and bus data can be retrieved for the development of LabVIEW application for instrument control. In Chap. 4, you will see that GPIB read and write functions are used to develop the attenuation measurement suite on LabVIEW virtual instrument.

Fig. 2.5 GPIB functions
a read and b write

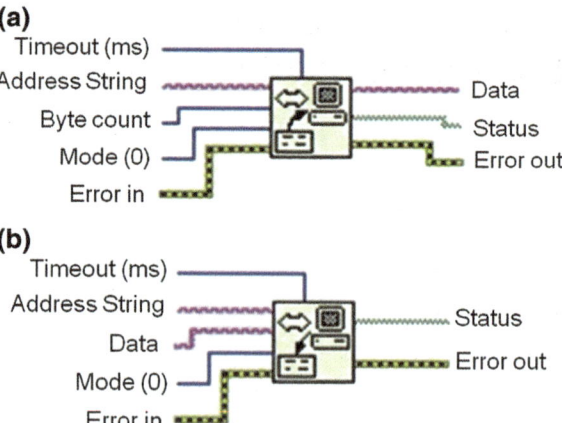

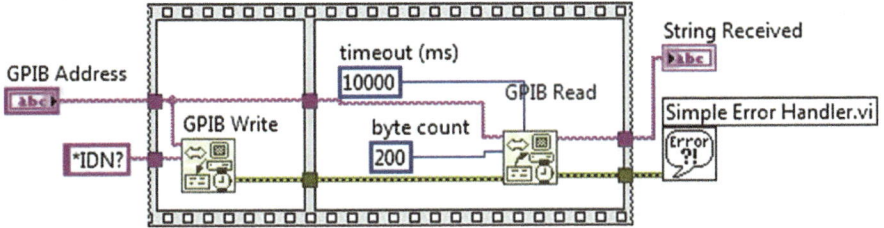

Fig. 2.6 Simple GPIB read and write functions used to communicate with the connected GPIB device

2.4 Advanced LabVIEW Programming Techniques

Although the programming capabilities of LabVIEW are large in number, only the techniques used in the subsequent chapters are discussed in this section. The major focus of this book is to provide the basic introduction to the programming techniques required to automate the commonly used microwave tests and measurements. Previously, the use of GPIB read and write function is demonstrated which plays an important role in GPIB instrument control. Additionally, the advanced file functions, dialogs, and user interface functions facilitate the development of user-friendly virtual instrument. The use of these advanced LabVIEW features will be demonstrated in Chap. 4.

References

1. R. Bitter, T. Mohiuddin, M. Nawrocki, *LabVIEW: Advanced programming techniques*, 2nd edn. (CRC Press, Florida, 2006). ISBN: 0-8493-3325-3
2. G.W. Johnson, *LabVIEW Graphical Programming*, 4th edn. (Tata McGraw-Hill Education, New York, 1997). ISBN: 0-07-145146-3
3. J. Travis, J. Kring, *LabVIEW for Everyone: Graphical Programming Made Easy and Fun (National instruments virtual instrumentation series)*, 3rd edn. (Prentice Hall PTR, New Jersey, 2006). ISBN: 0131856723
4. H. Goldberg, What is virtual instrumentation? IEEE Instrum. Measur. Mag. **3**(4), 10–13 (2000)

Chapter 3
Microwave Measurement Systems

In this chapter, the measurement of common microwave parameters such as RF power, attenuation, and scattering (S-) parameters are explained. An example of automated attenuation measurement is taken in detail in the next chapter based on the understanding acquired through this chapter about the nature of microwave measurements. A brief description of each parameter, instrumentation, its measurement setup, and uncertainty evaluation is given in the chapter. The first section introduces the reader to the basic measurement equipment used in modern microwave measurements. The next section enlists the common microwave measurements. The microwave measurement methods are chosen and explained in order of increasing complexity. For example, the power ratio technique uses a simple and generally available power meter and sensor, whereas there are also systems available that are specific to a single parameter, for instance, TEGAM's VM-7 for microwave attenuation measurement. But some systems are even more versatile, such as spectrum analyzer and vector network analyzer (VNA), which are used in almost all microwave measurements.

Attenuation is one the most important properties of the device under test (DUT), explained in the subsequent text. It is hence chosen for the worked out example of automation in next chapter. The RF power measurement is also used for attenuation measurement, which is based on *Power Ratio* principle. Subsequently, the understanding of microwave measurements will ensure readers' ability to understand the automation process.

3.1 Basic Microwave Instruments

Depending on the type of the measurement, generator and receivers vary from one setup to another. For example, if attenuation is being measured for a fixed or step attenuator, then the source can be a signal generator which may or may not be a synthesized source. Similarly, the receiver varies. For attenuation measurement,

© The Author(s) 2018
S. Dubey et al., *LabVIEW Based Automation Guide for Microwave Measurements*,
SpringerBriefs in Computational Electromagnetics,
https://doi.org/10.1007/978-981-10-6280-3_3

spectrum analyzer or VM-7 is used as a receiver. On the other hand, for S-parameter measurement VNA is used as a source as well as the receiver. Generally, software support for VNA is provided by the instrument manufacturers itself for data acquisition. However, customized solutions can be developed as explained in the subsequent chapters.

3.1.1 Microwave Signal Generators

Microwave signal generator is an electronic device that is used to generate the microwave signal in the analog or digital domain. It is used as the source for microwave measurements. Figure 3.1 enlists the various type of signal generators commonly used in testing and measurements.

Function generators are suitable when different waveforms are to be studied. In today's scenario, free running signal generators have become obsolete due to the better performance of synthesized sources. On the other hand, vector network analyzer is a complete package. The source in VNA may be an oscillator or synthesized source.

3.1.2 Microwave Receivers

The detection of a signal at the receiver end is important for any measurement. The detection techniques are evolving day by day. Some of the commonly used receivers in microwave tests and measurements are shown in Fig. 3.2.

(a) *Power Sensor and Meter*:

There are different techniques to measure microwave power. Based on the user requirements, the techniques are adopted. The three main methods are given in Fig. 3.3. Thermocouple- and thermistor-based power sensors are usually used for higher accuracy, linearity, and stability. Diode detectors are widely used otherwise where high accuracy is not essential.

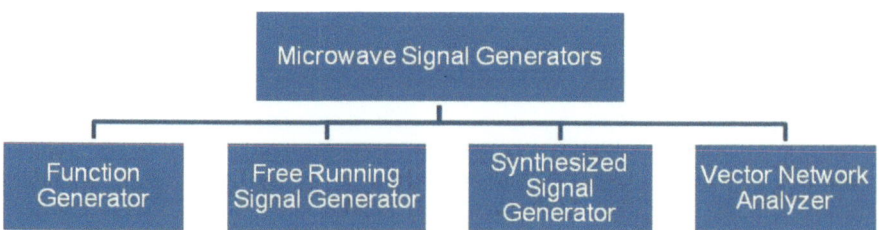

Fig. 3.1 Commonly used microwave signal generators

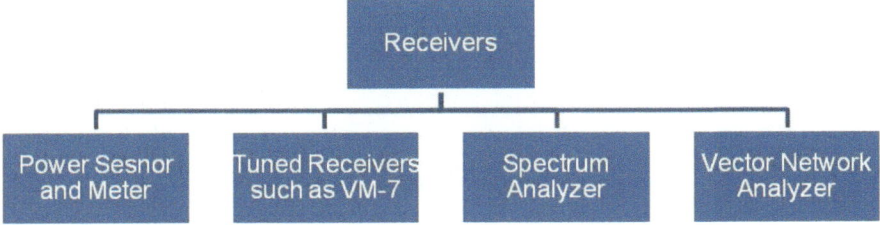

Fig. 3.2 Commonly used receivers

Fig. 3.3 Types of power sensors

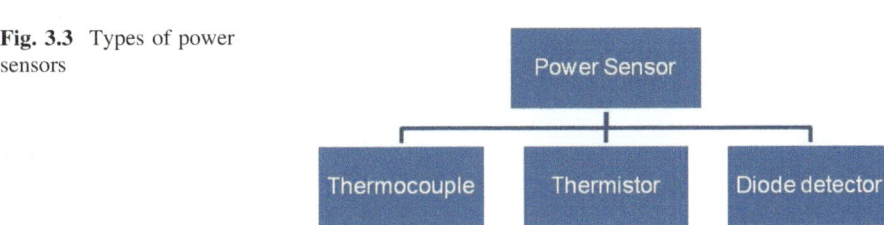

(b) *Tuned Receivers*:

A tuned receiver is used for high sensitivity and dynamic range. For example, commercially available TEGAM's VM-7 is a 30 MHz tuned receiver which is used with a mixer to down-convert the input RF signal to generate a 30 MHz intermediate frequency (IF). A local oscillator is present in VM-7 which is locked with the generated 30 MHz signal for higher sensitivity.

These receivers tuned to a single frequency may play an important role in the precise measurements of various parameters such as insertion loss, attenuation, characteristics of microwave amplifiers or mixers, antenna measurement and RF leakage.

(c) *Spectrum Analyzer*:

A spectrum analyzer is a measuring device used to display the signal in frequency domain. It is used for various applications such as measuring modulation, noise, harmonic and inter-modulation distortion. Apart from studying the properties of the signal, electromagnetic compatibility diagnosis can also be made using spectrum analyzers. In a way, the display of the spectrum analyzer gives the physical realization of the signal and hence is widely used in microwave measurements.

(d) *Vector Network Analyzers*:

Vector network analyzers (VNAs) are capable of providing magnitude as well as the phase information for the measured S-parameter. The receiver in the VNA is now generally a tuned receiver. However, in some cases, diode detection is also used at the receiver end of the VNA, but drift in diode detector limits its performance against the tuned receiver which has the better dynamic range and sensitivity. For further details on the microwave devices, one may refer to Carvalho and Schreurs [1].

3.2 Basic Microwave Measurement Methods

Once the reader is acquainted with the microwave measurement devices, it may get easier to understand the measurement of different microwave parameters (Fig. 3.4).

3.2.1 Attenuation Measurement

Basic Concepts: Attenuation is a pure property of the device. It can be understood as an insertion loss generated when a device is inserted in a perfectly matched condition between the generator and load. It can be measured as the change in the indicated power when RF power from an impedance matched source is passed first directly and secondly through the DUT into a matched power sensor. The power from the sensor is measured by a power meter. Attenuation is then expressed as

$$A(\text{dB}) = 10 \log_{10}\left(\frac{P1}{P2}\right), \tag{3.1}$$

where $P1$ is the power indication without the attenuator in line, and $P2$ is the power indication with the attenuator in line. The basic measurement setup is shown in Fig. 3.5.

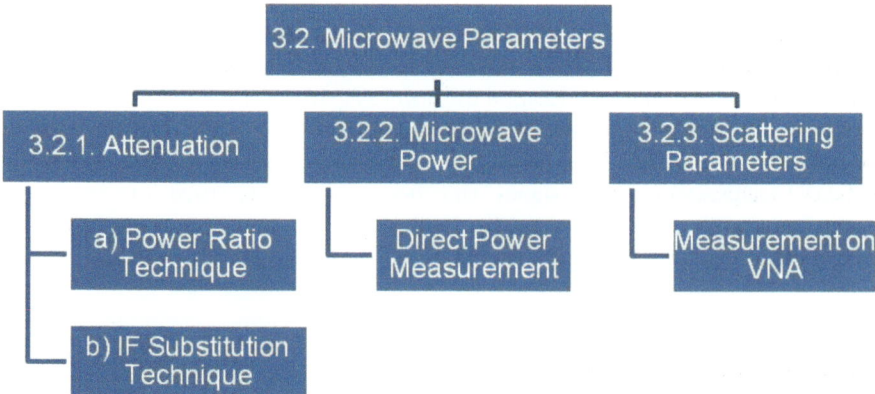

Fig. 3.4 The essential microwave parameters and their measurement discussed in this chapter

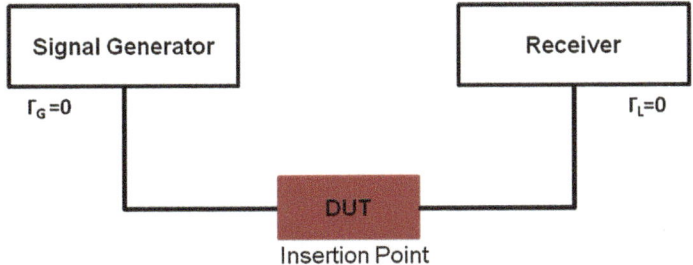

Fig. 3.5 Attenuation measurement

The main techniques used for microwave attenuation measurement are power ratio technique and IF substitution technique.

(a) *Power Ratio Technique*:

In Fig. 3.5, the receiver can be a power meter, VM-7 or a spectrum analyzer depending on the measurement principle being implemented. If power ratio technique is being used to measure attenuation, then the receiver will be a power meter, shown in Fig. 3.6. It is perhaps one of the easiest to configure. In power ratio technique, the power sensor is preceded by a matching attenuator pad to an RF source followed by a power sensor and the power meter indication is noted as $P1$. Next, the device under test is inserted between the matching pad and power sensor. Again the power meter indication is noted say, $P2$. The insertion loss is then calculated using Eq. (3.1). Note that, unless the reflection coefficient of the

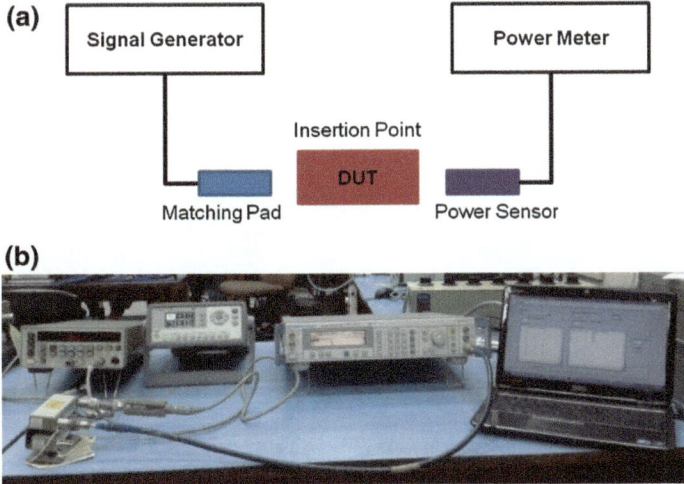

Fig. 3.6 Power ratio technique **a** Schematic Diagram. **b** Attenuation measurement system controlled by the PC using LabVIEW application

generator (Γ_G) and load (Γ_L) at the insertion point is known to be zero, in Fig. 3.5, or that the mismatch factor has been calculated and taken into consideration, measured insertion loss and no attenuation is quoted. It is because the attenuation is the property of the two-port network, whereas it is insertion loss which considers the mismatch losses of signal generator and load (Γ_G and Γ_L).

(b) *IF Substitution Technique*:

If Intermediate Frequency (IF) substitution technique is implemented, then the basic measurement system shown in Fig. 3.5 modifies as shown in Fig. 3.7. In the IF substitution technique, the system compares the attenuation through the device under calibration with an IF attenuation standard. A mixer along with an IF receiver is introduced into the measurement system to generate the desired IF frequency. To perform this substitution method, two different RF sources are required. A mixer with calculated nonlinearity is deployed between the RF source and Local oscillator. The mixer is a three-port device, two RF source will act as an input of the mixer, and the difference between these frequencies will generate an intermediate frequency on the output port. In the measurement, 30-MHz substitution technique is deployed because of the 30-MHz tuned receiver (VM7), shown in Fig. 3.7. Here, VM7 is a 30-MHz tuned receiver with the in-built local oscillator.

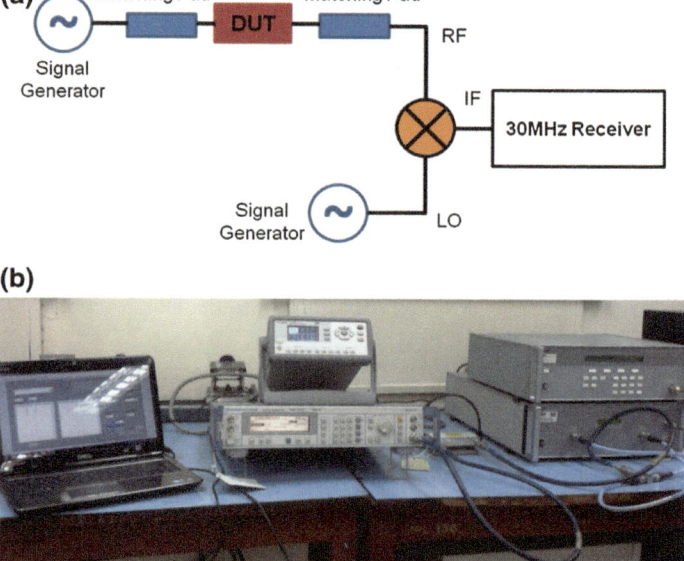

Fig. 3.7 30 MHz IF substitution technique **a** Schematic diagram. **b** Attenuation measurement system controlled by the PC using LabVIEW application

3.2.2 *Microwave Power Measurement*

In the RF or microwave frequency range, the power is the measure of signal strength. It is given in Watts or dBm. The power in dBm denotes the logarithmic value of 1 mW power, i.e., given as

$$P_{\text{dBm}} = \log_{10}\left(\frac{P}{1\ \text{mW}}\right). \tag{3.2}$$

Microwave power is measured using power sensors. The source is connected to the load via power sensor, shown in Fig. 3.8. Power sensors are basically used for converting the microwave power into measurable DC signal. There are mainly three types of sensors—thermistor based, thermocouple based, and diode detector based. Power measurement is of great importance in different fields such as telecommunication and testing of equipments.

With recent introduction of systems, such as Fluke 96270A 27 GHz Low Phase Noise Reference Source (Fig. 3.9), it is observed that performance of the power sensors are improving. With the advent of technology, newer and improved power sensors are making their place in the market due to their lower mismatch error, higher linearity, and wider range as compared to RF mixers. Wideband power sensors based on RF MEMS technology have been realized [2], and thus it is expected that RF systems will be using power sensors soon to replace the present market of resistive attenuators.

3.2.3 *Scattering Parameter*

Scattering parameters define the forward and reverse wave propagation through a network. Abundant literature is available on the theoretical elucidation of S-parameters [3]. A brief description here will help in understanding the basic concept of S-parameter and its measurement. Considering the example of a two-port network, shown in Fig. 3.10, S-parameters should define input and output from the network. In general, the S-parameters for a two-port network can be given as

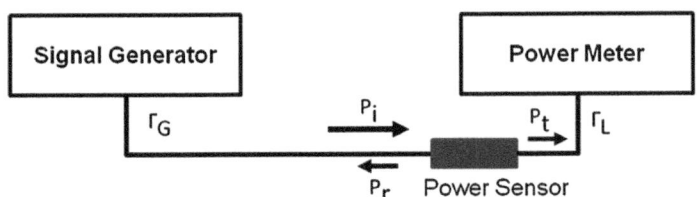

Fig. 3.8 Power measurement system

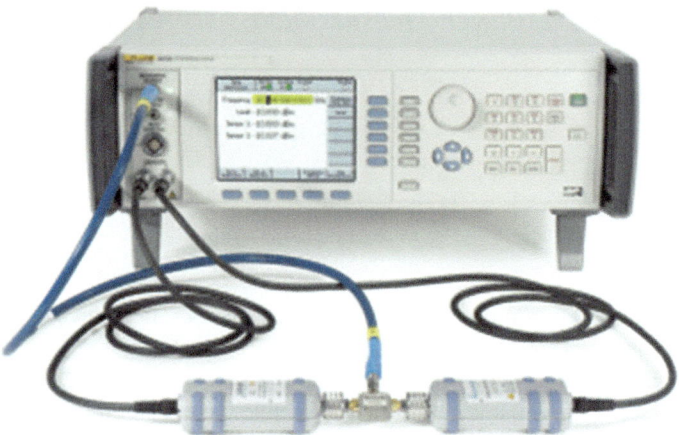

Fig. 3.9 An example of RF system that exploits the linearity advantage of power sensors—Fluke 96270A 27 GHz low phase noise reference source. *Courtesy* Fluke Calibration

Fig. 3.10 Two-port device in terms of S-parameters

$$b_1 = S_{11}a_1 + S_{12}a_2, \tag{3.3}$$

$$b_2 = S_{21}a_1 + S_{22}a_2, \tag{3.4}$$

where a_1 and a_2 are the signals entering the Port 1 and Port 2 of the two-port network and b_1 and b_2 are the signals leaving the respective ports.

So, for characterizing a two-port network for its S-parameters, a VNA comprising of a RF source, bridges/couplers, mixers, and tuned receiver are required in order to measure the forward and reverse signals at both the ports.

3.3   Uncertainty Evaluation in Attenuation Measurement

Nowadays, it is a general practise to evaluate the uncertainty involved in the experiment. It is unlike the previous days, when only metrologists working in the field of precision measurements used to evaluate the uncertainty components of the experiment. Besides, the meaning of uncertainty should be understood carefully. The *"Guide to the Expression of Uncertainty in Measurement"* (GUM) [4] defines uncertainty as a "parameter, associated with the result of a measurement that characterizes the dispersion of the values that could reasonably be attributed to the measurand."

However, the uncertainty should not be confused with error in measurement. The error is difference between the measured value and the true value of the measurand. So, for example, if reading 9.995 dB attenuation from a 10 dB (Nominal value) attenuator, then 0.005 dB (= (10 − 9.995) dB) is the error. On the other hand, the uncertainty in the measurement is to be calculated from the doubtful parameters (e.g., leakage, cable loss, etc.). For example, a result given as 9.995 ± 0.010 dB, where 9.995 dB is the measured value and ±0.010 dB is the uncertainty of the result. In this result, all the doubts have been quantified and included in the result as ±0.010 dB.

The uncertainty contributions are analyzed to show confidence in the measurement results. The GUM document compiles the procedure to calculate the uncertainty. Adherence to the given procedure in GUM document creates uniform understanding between different research laboratories all over the world. GUM document is applicable for a large number of measurements and can be used accordingly. National Measurement Institutes (NMIs) and accreditation laboratories must use GUM document for uniform understanding and expression of uncertainty [5].

When uncertainty is assessed, generally, some parameters contribute to uncertainty of wide number of measurements, for example, meter resolution. Every measurement consists of a receiver on which the measurand is read. And that very receiver will have a limit on the resolution. A receiver may be capable of reading 0.1 dB attenuation, whereas other may possibly read up to 0.001 dB value. If the limit is 0.1 dB, then we are definitely doubtful about the 1/100th place. And hence it contributes to the uncertainty in the measurement.

There are some other parameters which are specifically the uncertainty contributors in microwave measurements such as connector repeatability, mismatch losses, directivity, and linearity. Here, the main concerns are the errors that are caused by the RF leakage, power instability, frequency drift, and RF connectors.

Once the uncertainty contributors are found and measured. The value of each contributor is used to calculate the combined uncertainty (U_c). GUM defines combined uncertainty as "standard uncertainty of the result of a measurement when that result is obtained from the values of a number of other quantities, equal to the positive square root of a sum of terms, the terms being the variances or co-variances of these other quantities weighted according to how the measurement result varies with changes in these quantities." The combined uncertainty (U_C) can be defined as

$$U_C = \sqrt{U_A^2 + U_{B1}^2 + \cdots + U_{Bn}^2},\qquad (3.5)$$

where U_A is the Type-A and $U_{B1} \ldots U_{Bn}$ are the Type-B uncertainty contributors [4]. And the expanded uncertainty (U), defined as a "quantity defining an interval about the result of a measurement that may be expected to encompass a large fraction of the distribution of values that could reasonably be attributed to the measurand", is given by equation

$$U = k \times U_C.]] > \qquad (3.6)$$

k is calculated from student's t-distribution. The value of k unambiguously states the level of confidence associated with the uncertainty interval. For example, $k \approx 2$ has been used that produces an interval having a level of confidence of approximately 95%. Hence, if measured attenuation is 9.995 ± 0.010 dB at confidence level of 95%, then we are 95% sure that attenuation value is between 9.985 and 10.005 dB.

3.3.1 Uncertainty in Power Ratio Technique

Apart from these errors, the error in the power ratio technique includes the power sensor linearity. The Power Ratio Technique is simple and easy to configure, but it has some limitations, such as amplitude drift, which is directly proportional to the signal generator output amplitude drift, zero carry over, and power linearity of the power sensor. The power sensor linearity is a key source of error in power ratio technique as per mismatch uncertainty is concerned. Diode power sensors, described by Cherry et al. [6], may be modeled by the following equation:

$$P_{in} = xV_{dc}\exp(yV_{dc}), \qquad (3.7)$$

where P_{in} is the incident power, V_{dc} is the rectified DC voltage, and x and y are the constants which are functions of parameters such as temperature, ideality factor, and video impedance. However, power sensor linearity can be improved as given in [7].

3.3.2   Uncertainty in IF Substitution Technique

The mixer nonlinearity is the main uncertainty contributor in IF substitution technique [8]. It is worth noting that the IF substitution technique uses two RF sources and its counterpart, the power ratio technique, uses a single source. The power ratio technique hence has lower error due to frequency and power instability. In addition, the errors introduced due to mismatch of the DUT and attenuation standard remains same for both the techniques. Mismatch can be simply defined as the difference between the insertion loss and attenuation. For our particular work, the attenuator is a variable attenuator and its mismatch error is calculated by the formula [8]

$$
\begin{aligned}
M(\text{dB}) = 20 \log_{10}[1 &- \Gamma_G(\Gamma_{I1} - \Gamma_{I2}) - \Gamma_D(\Gamma_{O1} - \Gamma_{O2}) \\
&+ \Gamma_G \Gamma_D \left(10^{-A1/10} - 10^{-A2/10} \right)]
\end{aligned}
\tag{3.8}
$$

where Γ_G, Γ_D are the reflection coefficient of the source and the detector, Γ_{I1}, Γ_{I2} are the input reflection coefficient of the DUT for two settings of attenuation, Γ_{O1}, Γ_{O2} are the output reflection coefficient for two settings of attenuation and $A1$, $A2$ are the attenuation values of the DUT in the two settings.

3.3.3   Comparison of Measurement Results

After the attenuation measurement and uncertainty evaluation in manual and automated experiments, our aim is to compare the results. In order to compare the quality of automated measurement against the manual measurement, the conventional approach taken during the inter-laboratory key comparisons can be used. Hence, if the uncertainty is estimated for both the techniques, then the quality of automated measurement can be judged against the manual measurement using investigation on normalized error, E_n. The E_n-ratio is given by

$$
E_n = \frac{\text{Lab} - \text{Ref}}{\sqrt{U_{\text{Lab}}^2 + U_{\text{Ref}}^2}},
\tag{3.9}
$$

where Lab is the measurement value, Ref is the reference value, U_{Lab} is the uncertainty reported by the technique, and U_{Ref} is the total uncertainty of the reference value (including any allowance for drift or instability of the DUT). The values of $|E_n| > 1$ require investigation. Where laboratories make a number of similar measurements the method of analysis can be refined by comparing the distribution of the values of E_n with a normal distribution [9].

3.4   Summary

The chapter discusses the important microwave measurements and its uncertainty evaluation. The major parameters such as attenuation, microwave power, and scattering parameters are introduced and its metrological aspects are discussed. A generalized method in metrology of such parameters is depicted in Fig. 3.11.

Fig. 3.11 Generalized method for microwave measurement

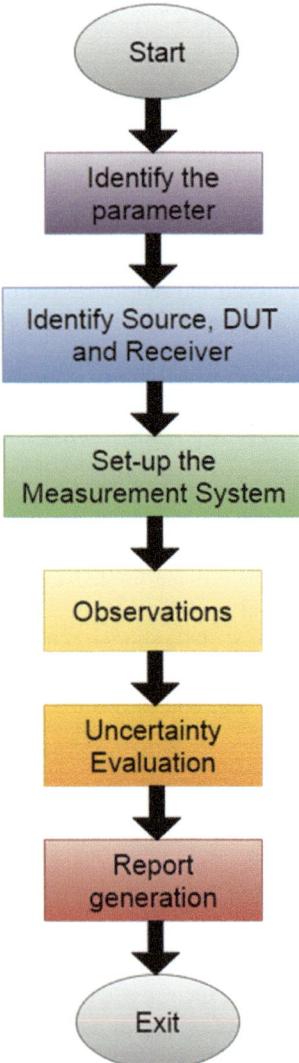

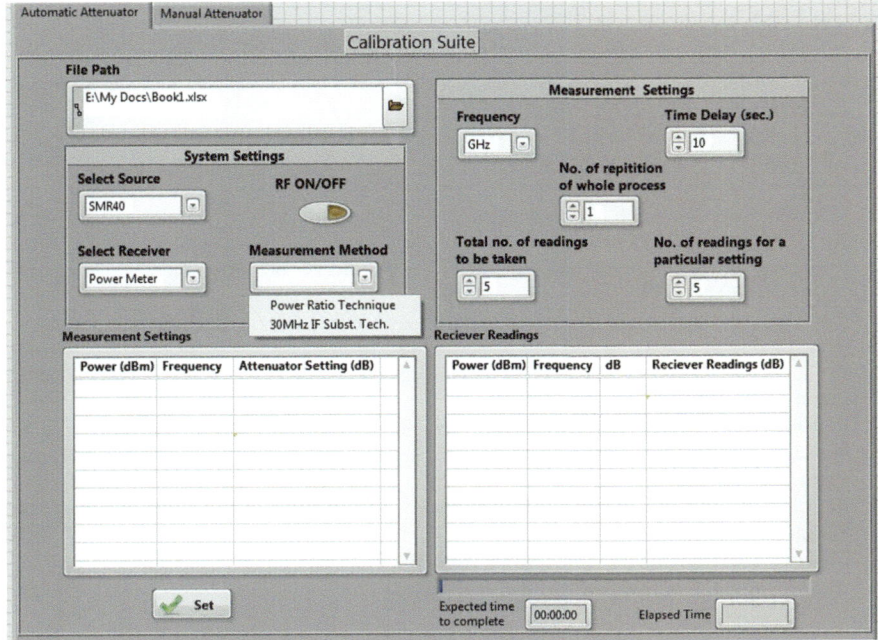

Fig. 3.12 Custom designed LabVIEW application for attenuation measurement using power ratio or IF substitution technique

3.5   Precap

On understanding the above-mentioned measurement techniques, we can now classify the instruments into categories of their operation. To develop the software solution for the measurement systems, we now have the basic understanding of LabVIEW programming given in Chap. 2. This chapter discussed the main consideration for microwave measurements and created understanding of the flow of control to be simulated in the software. Chapter 4 ahead will discuss the software development techniques required for building your own software solution for microwave measurement systems. Figure 3.12 shows the graphical user interface of the software that can be developed by the reader for various measurement methods as the outcome of understanding of this book.

References

1. N.B. Carvalho, D. Schreurs, *Microwave and Wireless Measurement Techniques* (Cambridge University Press, New York, 2013). ISBN: 1107004616
2. L.J. Fernández, R.J. Wiegerink, J. Flokstra, J. Sesé, H.V. Jansen, M. Elwenspoek, A capacitive RF power sensor based on MEMS technology. J. Micromech. Microeng. **16**(7), 1099–1107 (2006)
3. R.A. Witte, *Spectrum and Network Measurements*, 2nd edn. (SciTech Publishing, New Jersey, 2014). ISBN: 9781613530368
4. Evaluation of measurement data—guide to the expression of uncertainty in measurement. JCGM 100:2008, BIPM 2008
5. V.N. Ojha, Evaluation and expression of uncertainty in measurement. MAPAN J. Metrol. Soc. India **13**, 71–84 (1998)
6. P. Cherry, W. Oram, G. Hjipieris, A dynamic calibrator for detector nonlinearity character-ization. Microw. Eng. Europe (1995)
7. K. Holland, J. Howes, Improvements to the microwave mixer and power sensor linearity measurement capability at the national physical laboratory. IEE Proc. Sci. Measur. Technol. **149**(6), 329–332 (2002)
8. R. Collier, D. Skinner, *Microwave Measurements*, 3rd edn. (The Institution of Engineering and Technology, London, 2007)
9. Calibration interlaboratory comparisons. APLAC. **5**, 1–23

Chapter 4
Software Development

Software for metrology is required for the reliable, fast, and improved test and measurement results. Software development for metrology needs details about the standards and the measurement process. Chapter 3 has detailed the major measurements in microwave engineering. Now that the reader has the detailed information about the measurements in microwave, let us implement the measurement system using LabVIEW for attenuation measurement as a hands-on exercise. A step-by-step approach detailed in this chapter for implementation of attenuation measurement system. However, the understanding gained from this chapter can be used for any other measurement that uses GPIB devices, mainly the source, DUT, and the receiver.

4.1 LabVIEW-Based Automatic Measurement System

A software unit is developed on LabVIEW platform and is used for attenuation measurement techniques in order to acquire measurement data, evaluate uncertainty, and performing other statistical analyses required before reporting the measurement results. The application is validated in order to improve and maintain the quality of metrological performance. The crucial guidelines and considerations for validating the software performance are widely discussed in the standard documents. The major guidelines of ISO/IEC: 17025 [1] are followed during the design and implementation of the software. ISO/IEC: 17025 emphasizes on the validation of the user-developed software, when used for the acquisition, processing, storage or retrieval of test, or calibration data. The performance of software thus is evaluated by comparing the manual results with the automated ones along with the considerations that were made during the design process.

The LabVIEW application is created using simple message-based programming where the GPIB address uniquely identifies the instrument, and GPIB Read/Write functions are used to communicate with the instrument. The front panel provides

© The Author(s) 2018   25
S. Dubey et al., *LabVIEW Based Automation Guide for Microwave Measurements*,
SpringerBriefs in Computational Electromagnetics,
https://doi.org/10.1007/978-981-10-6280-3_4

the basic features required to initiate the measurement experiment, specifically communicating with the connected instruments to set them as per the system settings given by the user.

4.1.1 Software Requirement Specifications

The software requirement specification (SRS) provides the description of the requirements to be fulfilled by the proposed software [2]. The requirements specified in the SRS document covers the functional as well as nonfunctional requirements. The SRS clearly defines the software functionality, Software–Hardware interfaces, user interface, system risk assessment and validity, performance criteria, and implementation issues. The basic functions of metrology software are controlling, monitoring, displaying, and reporting the results of the measurement process.

For example, Fig. 4.1 shows an example of software requirements for attenuation measurement software. Once the software requirements are known, the design and implementation of the software should be started fulfilling the requirements of the customer/end user.

4.1.2 Design of Graphical User Interface (GUI)

A graphical user interface (GUI) uses visual elements that provide the display for user input and output to and fro from the computer and human in an

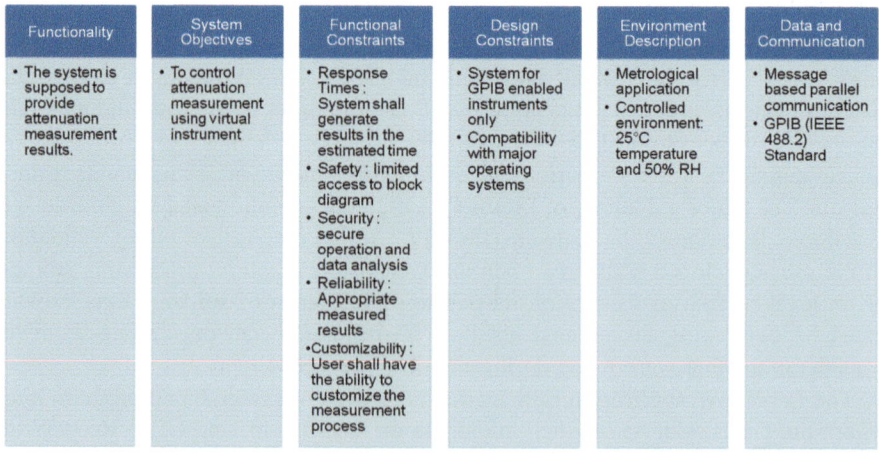

Fig. 4.1 An example of software requirement specification of attenuation measurement software

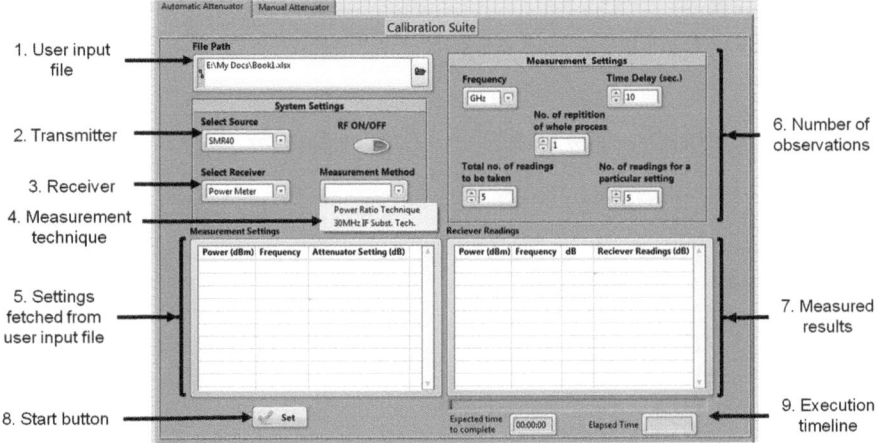

Fig. 4.2 Example GUI of the developed automated calibration software

easy-to-understand manner. Figure 4.2 shows the developed GUI for attenuation measurement. The controls and indicators on the front panel are labeled.

The following are major elements of the GUI:

1. User input file: When the application is run, it prompts for the file path, necessarily an excel sheet, in which the desired settings for measurement are recorded by the user.
2. Transmitter: The select source control on the front panel can be used to detect the GPIB source connected with the controller. One can provide the model of the source in the drop down list.
3. Receiver: The select receiver control on the front panel can be used to detect the GPIB enabled microwave receiver such as power meter or spectrum analyzer connected with the controller. One can provide the model of the receiver in the drop down list.
4. Measurement Technique: Based on the measurement technique, different data flow will be required.
5. Settings fetched from user input file: The measurement settings saved by the user in the MS Excel sheet can be fetched and shown in the table for real-time monitoring.
6. Number of observations: Once the user input is known, the time delay between the observations, number of observation set, total number of readings, etc., can be asked by the user under measurement settings.
7. Measured results: The receiver's readings against the user-defined settings can be seen on the GUI in the real time. This provision helps in the user to monitor the real-time readings.
8. Start button: A button to start the execution can be provided for the user to start the measurement process.

9. Execution timeline: The expected time to complete and elapsed time can be a
    good indicator for the ease of the user.

4.1.3 Configuring Transmitter, DUT, and Receiver

In a microwave measurement, as discussed in Chap. 3, the transmitter, DUT, and
receiver are to be controlled. For example, attenuation calibration software will
require controlling the instruments given in Table 4.1. It is to be noted that the
instrument-specific instructions can be found in the instrument user guides and
manuals [3–5]. These instructions can be sent and received to and fro from the
device using GPIB functions.

For example, see Fig. 4.3a, b to understand that GPIB write and read operations.
In Fig. 4.3a, SCPI command is sent to the signal generator for turning on the RF
source. In Fig. 4.3b, these functions are used to execute the query to measure
attenuation read by the GPIB device, in this case Tegam's VM-7, according to the
time delay defined by the user on the front panel.

4.1.4 System Architecture

The conceptual model for the automated microwave measurement system is shown
in Fig. 4.4. The end user will communicate with the controller through the GUI or
the VI front panel. The controls and indicators on the front panel are provided in
such a manner that the security and reliability of the software cannot be at risk by
the user. As per the SRS given above, the controller communicates over the GPIB
which was discussed in Chap. 2. The source code of the LabVIEW application lies
in the block diagram and is secured for better reliability.

System Hardware: The hardware of the automated system consists of

- IEEE-488.2 General Purpose Interface Bus (GPIB) kit,
- a microwave signal generator,

Table 4.1 Configuring instruments

Instrument	Model	Basic operations	Instructions/Queries
Source	R&S synthesized signal generator SMR40	RF on/off	OUTP:STAT ON
		Set frequency (e.g., to 2.45 GHz)	FREquation 2.45 GHz
		Set power level (e.g., to -10dBm power level)	POW-10 dBm
DUT	Agilent Programmable Attenuator 8494H	Setting the attenuation value on the step attenuator using switch controller	e.g., 1 dB = A1B234
Receiver	Tegam's VM-7	Enquiring the attenuation level read by the receiver	ATTN?

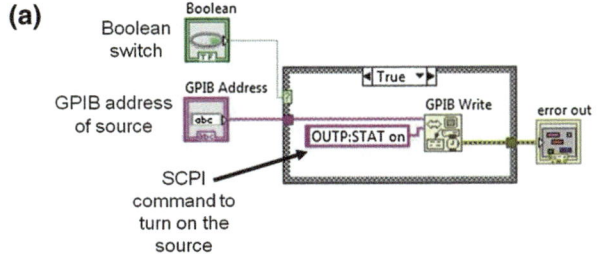

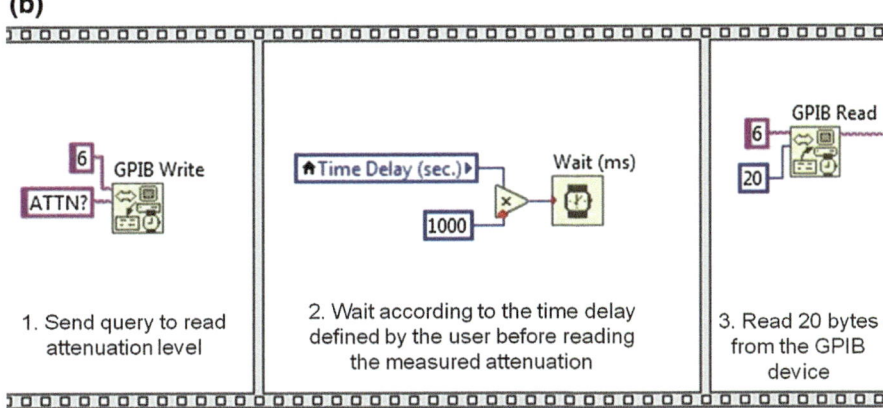

Fig. 4.3 Example of block diagram code for **a** turning on the source and **b** reading attenuation from Tegam's VM-7 (GPIB address: 6)

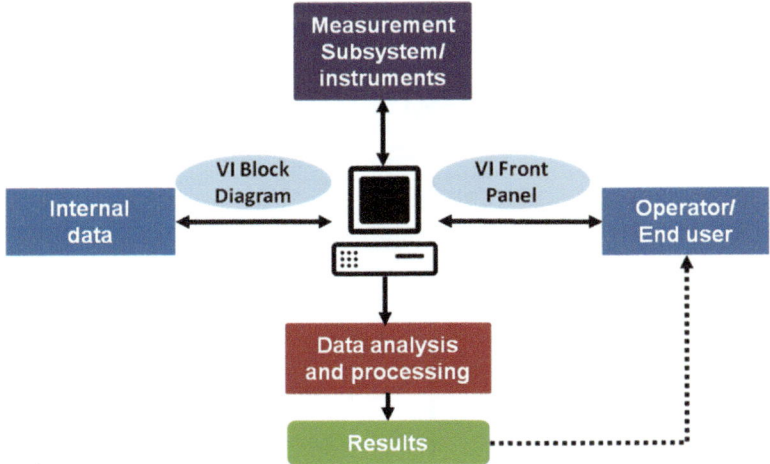

Fig. 4.4 A model of automated measurement system based on virtual instrumentation

- a programmable attenuator/switch driver,
- a step attenuator,
- a power sensor,
- a power meter,
- personal computer with NI LabVIEW software installed.

The instruments intended for control uses the star configuration in which the GPIB board connects with the Personal Computer (PC) via GPIB-to-USB adapter, the signal generator is connected to the GPIB board, and GPIB cable is then connected from the signal generator to the power meter (or VM-7 in case of IF substitution technique) and the attenuator driver. The flow of the GPIB Read and Write commands is schematically shown in Fig. 4.5.

4.1.5   System Implementation

The LabVIEW application is created using simple message-based programming where the GPIB address uniquely identifies the instrument and GPIB Read/Write functions are used to communicate with the instrument. The virtual instrumentation application has the front panel that works as a graphical user interface (GUI) and a block diagram at the back end.

The front panel provides the basic features required to initiate the measurement experiment, specifically communicating with the connected instruments to set them as per the system settings given by the user. When the application is run, it prompts for the file path, necessarily an excel sheet, in which the desired settings for measurement are recorded by the user as shown in Fig. 4.6.

Fig. 4.5 Flow of GPIB functions in power ratio technique

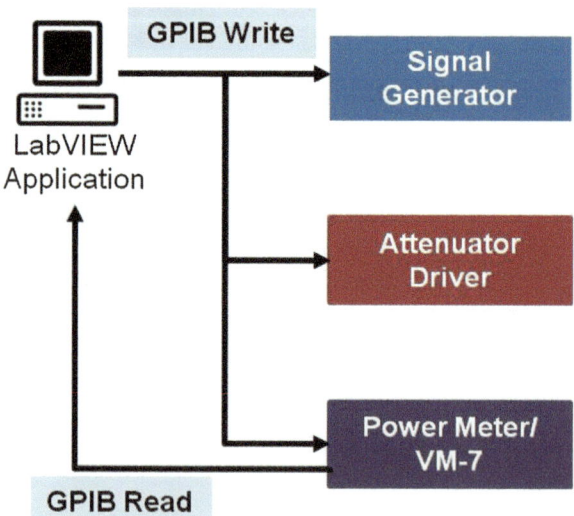

	A	B	C	D
1	**Power**	**Frequency**	**dB**	
2	0	2	0	
3	0	2	10	
4	0	2	0	
5	0	2	20	
6	0	2	0	
7	0	2	30	
8	0	2	0	
9	0	2	40	
10	0	2	0	
11	0	2	50	
12	0	2	0	
13	0	2	60	
14				
15				

Fig. 4.6 Simple measurement settings file to be given by the user using MS Excel

Table 4.2 Main LabVIEW Functions used in given example

Palette	Sub-Palette	Function	Used for
Instrument I/O	GPIB	GPIB read and write	Instrument control: GPIB Read and Write is used for reading instrument result (output) buffer and sending query to the instrument, respectively
Programming	Structures	For loop and while loop	Iteration/repetition
	Dialog and user interface	File and message dialog	User-friendly GUI
	Timing	Elapsed time	Progress bar
	Array	Array size, index array, insert into array, etc.	Data management

The main functions that can be used in LabVIEW application for instrument control are given in Table 4.2. These functions are of great importance to develop a user-friendly software. One can realize that these functions can ease the programmer from getting into too much of programming details. For example, the execution timeline on the front panel can be implemented easily as shown in Fig. 4.7.

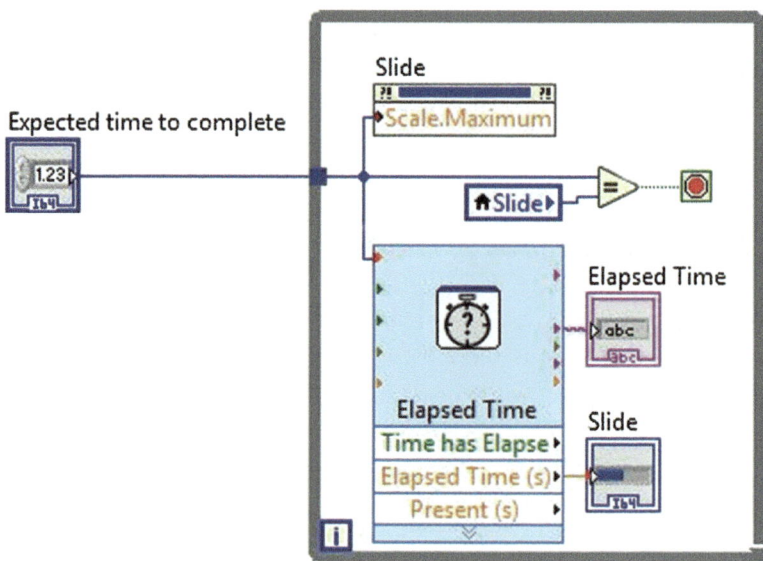

Fig. 4.7 VI for execution timeline

4.1.6 System Features and Drivers

Some features that add to the functionalities of the application are that the it is capable of dealing with fully programmable attenuators as well as the manual attenuators; application gives the real-time readings on the front panel as well as provides a complete observation sheet at the end of the experiment and time delay between consecutive readings can be set. Therefore, the application gives a standalone platform to deal with a number of instruments on a single screen with customizable settings and less manual intervention. The front panel of the LabVIEW application is shown in Fig. 4.2.

Solution Flexibility and Ease Compared to Previous Solutions

The developed application is compared with two previously used technologies for automation

(a) **Agilent VEE Pro**—It was found that LabVIEW development of software is easier and required lesser development time. Though VEE Pro is also a graphical language, but ease with the LabVIEW is better.
(b) **SureCal Calibration Software**—Developed LabVIEW application solved the problem of customized time interval between the readings.

4.2 Software Validation and Testing

There are a number of toolkits provided by NI LabVIEW for implementing automated software engineering phases. For refactoring our application to have a cleaner graphical code, the main toolkits we used upon the code are Unit Test Framework and VI Analyzer. These toolkits automate the validation phase of the software engineering model. The software application can be put under the unit testing with the LabVIEW Unit Test Framework toolkit in which test cases are run. Similarly, VI Analyzer is used for code review.

The use of software in metrology is required to be validated and assured for accurate and precise results [6, 7]. As specified in the standards such as ISO/IEC: 17025, the software is well documented, checks the operating conditions of the automated instruments before every execution and stores the measurement data securely to maintain the integrity of the calibration results.

Software integrity levels are defined regarding the complexity of the techniques to be used to validate the software. The following tests were performed on the software to validate the system:

(a) Code review
(b) Component Testing
(c) Regression Testing.

Hence, the software is validated up to the third level of software integrity. The software evaluates the uncertainty in the measurement as per the GUM document [8]. The validation is not only performed on the data acquisition module, but also for uncertainty calculation. The software's performance was found to be at par when evaluated on the guidelines available in the literature [1, 9–12].

References

1. P. Cherry, W. Oram, G. Hjipieris, A dynamic calibrator for detector nonlinearity characterization. Microw. Eng. Europe (1995)
2. K. Holland, J. Howes, Improvements to the microwave mixer and power sensor linearity measurement capability at the national physical laboratory. IEE Proc. Sci. Measur. Technol. **149**(6), 329–332 (2002)
3. R. Collier, D. Skinner, *Microwave Measurements*, 3rd edn. (The Institution of Engineering and Technology, London, 2007)
4. Calibration Interlaboratory Comparisons. APLAC **5**, 1–23 (2008)
5. ISO, BSEN. IEC 17025: 2005 General requirements for the competence of testing and calibration laboratories (2005)
6. ISO/IEC/IEEE International Standard—Systems and software engineering—Life cycle processes—Requirements engineering (2011), pp. 1–94. doi:10.1109/IEEESTD.2011.6146379
7. Operating Manual 1104.3430.12-01 Microwave Signal Generator. Rohde & Schwarz (1999)
8. H. Goldberg, What is virtual instrumentation? IEEE Instrum. Meas. Mag. **3**(4), 10–13 (2000)
9. .Operating and Service Manual HP438A Power Meter. Hewlett Packard (1986)

10. Agilent Technologies 11713A Attenuator/Switch Driver Operating and Service Manual. Agilent Technologies Inc. (2001)
11. T. Tasić, U. Grottker, An overview of guidance documents for software in metrological applications. Comput. Stand. Interfaces **28**(3), 256–269 (2006)
12. V. Batagelj, J. Bojkovski, J. Drnovšek, Software integration in national measurement-standards laboratories. IET Sci. Meas. Technol. **2**(2), 100–106 (2008)

Chapter 5
Measurement Results

Metrology is the science of measurement and its application. It may include both experimental and theoretical determinations in any field of science and technology. Broadly, the metrology is categorized into scientific, legal, or industrial metrology depending upon the type of application and level of accuracy and precision of the carried out measurement. In Chap. 3, details of some of the common microwave measurement were given, out of which automation of test measurement of attenuation is explained in Chap. 4. The proposed LabVIEW application can measure the attenuation of the DUT. The expression of the measurement results along with the evaluated uncertainty is explained in this chapter.

5.1 Data Analysis

The measured data, for example, the attenuation measured using power ratio of IF substitution technique can be analyzed using LabVIEW. The Mathematics palette of LabVIEW has various probabilistic and statistical functions that can be used for analyzing the measurement results. Figure 5.1 shows the complete idea behind the automated test setup of attenuation measurement. Once the measurement results are fetched and shown on the LabVIEW front panel using GPIB read function, the measured results can be analyzed, and uncertainty in the measurement can be evaluated. In the subsequent section, an example for uncertainty calculation using Mathematics functions is shown for better understanding.

© The Author(s) 2018
S. Dubey et al., *LabVIEW Based Automation Guide for Microwave Measurements*,
SpringerBriefs in Computational Electromagnetics,
https://doi.org/10.1007/978-981-10-6280-3_5

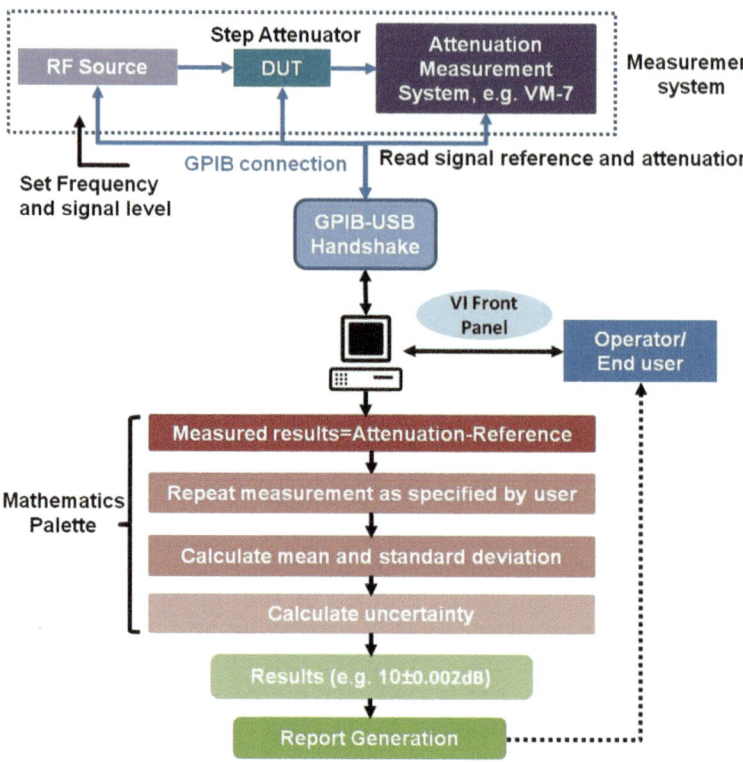

Fig. 5.1 Basic architecture for automated attenuation measurement setup

5.2   Uncertainty Evaluation

The key sources of uncertainties are identified and listed along with their probability distributions and limits, at the highest attenuation levels that can be measured, by power ratio technique and IF substitution technique in Tables 5.1 and 5.2, respectively.

Table 5.3 compares the results of manual and automated power ratio technique. It is interesting to note that, the manual and automated measurement shows a difference greater than their expanded uncertainties at 10 GHz for 10 dB attenuation. This is because the nonlinearity of the power sensor increases with the increasing frequency and attenuation values. Hence, for manual measurement, we use microwave tuners to obtain better matching condition, and in the case of automated measurement, fixed attenuators of 3 and 6 dB have been used as matching pads. Thus, at higher frequencies and higher attenuation levels, the difference between the automated and manual setup readings can be large but can be reduced by using automated microwave tuners (Fig. 5.2).

Table 5.1 Sources of uncertainty in power ratio technique up to 30 dB

Source of uncertainty	Limits ($\pm$dB)	Probability distribution
Power sensor linearity, U_{B1}	0.01	Rectangular
Meter resolution, U_{B2}	0.01	Rectangular
Mismatch uncertainty, U_{B3}	0.002	U-shaped
Repeatability, U_A	0.0045	Normal

Table 5.2 Sources of uncertainty in IF substitution technique up to 60 dB

Sources of uncertainty	Limits ($\pm$dB)	Probability distribution	Sensitivity coefficient
Repeatability, U_A	0.00837	Normal	1.0
Standard attenuator, U_{B1}	0.04	Normal	1.0
Meter resolution, U_{B2}	0.005	Rectangular	1.0
Mismatch uncertainty, U_{B3}	0.00682	U-shaped	1.0
Mixer nonlinearity, U_{B4}	0.0602	Rectangular	1.0
Isolation, U_{B5}	0.0282	Rectangular	1.0

Table 5.3 compares the results of manual (a) and automated power ratio (b) and 30MHz IF substitution (c) technique. From figures is quite evident that the two measurement techniques differ in the measurand value. It is to be noted that the two techniques are traceable to the two different reference standards, i.e., the thermistor mount and the WBCO attenuator. So, if the results of the two techniques differ with large values, then the evaluation of normalized error with respect to the reference values becomes inevitable.

Hence, for the validation of measurement techniques, the E_n ratio is evaluated by the LabVIEW application and found to be $|E_n| < 1$. Especially at frequency 10 GHz, attenuation measurement technique shows a difference greater than their expanded uncertainty at 5 dB and 10 dB. In both cases $|E_n| > 1$, hence their reference standard values have been considered to evaluate $|E_n|$. The front panel of the LabVIEW application will inform the user about E_n ratio acceptance with green and rejection with red. If there is a red indication, then a new dialog box will be executed and will ask for reference standard values in the form of MS Excel or comma separated values (CSV) file. The application will compute the E_n ratio between the measured values and the reference standard values. Hence, if finally $|E_n| < 1$ with respect to the reference, then the LED turns to green else a red indication will show that the measurement results are not valid, as illustrated in Fig. 5.3. Therefore, it is concluded that the quality of the measured results is not affected by the automation. Similarly, the values of the measurand using power ratio technique are traceable to the microcalorimeter-based thermistor mount standard for microwave power and should agree with that reference standard.

Table 5.3 Experimental results and comparisons

Frequency (GHz)	Nominal attenuation (dB)	Manual power ratio technique		Automated power ratio technique		Difference (a–b)	30 MHz IF substitution technique		Normalized error (E_n Ratio)	Difference (c–b)
		Measured attenuation (dB) (a)	Expanded uncertainty (±)	Measured attenuation (dB) (b)	Expanded uncertainty (±)		Measured attenuation (dB) (c)	Expanded uncertainty (±)		
0.03	1	1.007	0.008	1.006	0.007	0.001	1.002	0.011	0.460	−0.004
	5	5.030	0.012	5.031	0.010	−0.001	5.020	0.012	0.704	−0.011
	10	9.995	0.019	9.993	0.014	0.002	10.01	0.013	0.889	0.017
0.06	1	1.000	0.008	0.999	0.007	0.001	1.002	0.011	0.230	0.003
	5	5.009	0.012	5.008	0.010	0.001	5.010	0.015	0.443	0.002
	10	9.960	0.015	9.962	0.013	−0.002	9.968	0.016	0.776	0.006
1	1	1.001	0.007	1.001	0.007	0.000	1.012	0.010	0.901	0.011
	5	5.012	0.012	5.011	0.009	0.001	5.030	0.014	0.540	0.019
	10	9.967	0.015	9.966	0.013	0.001	10.016	0.016	0.970	0.050
2	1	1.001	0.007	1.001	0.007	0.000	1.016	0.012	0.360	0.015
	5	5.011	0.011	5.009	0.009	0.002	5.040	0.014	0.540	0.031
	10	9.966	0.015	9.964	0.013	0.002	10.024	0.016	0.970	0.060
5	1	1.005	0.007	1.004	0.007	0.001	1.028	0.012	0.288	0.024
	5	5.024	0.009	5.026	0.009	−0.002	5.076	0.014	0.601	0.050
	10	9.986	0.012	9.991	0.013	−0.005	10.060	0.020	0.377	0.069
10	1	0.945	0.007	0.969	0.007	−0.024	1.002	0.010	0.245	0.033
	5	5.044	0.009	5.068	0.010	−0.024	5.012	0.013	0.975	−0.056
	10	10.015	0.012	10.091	0.014	−0.076	10.032	0.017	0.862	−0.059

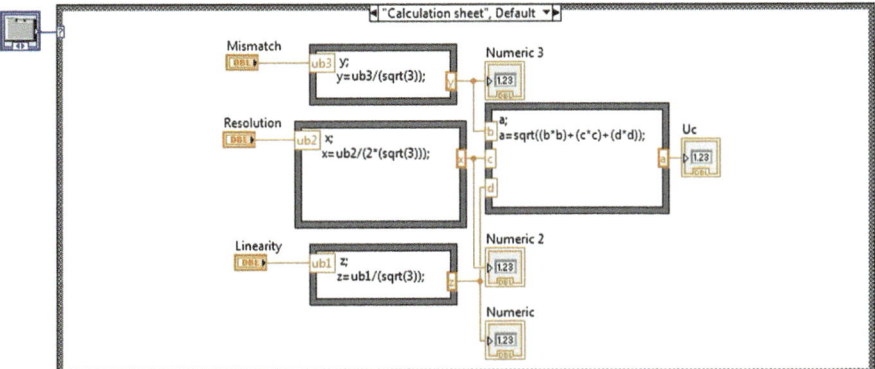

Fig. 5.2 Example LabVIEW VI block diagram for combined uncertainty evaluation

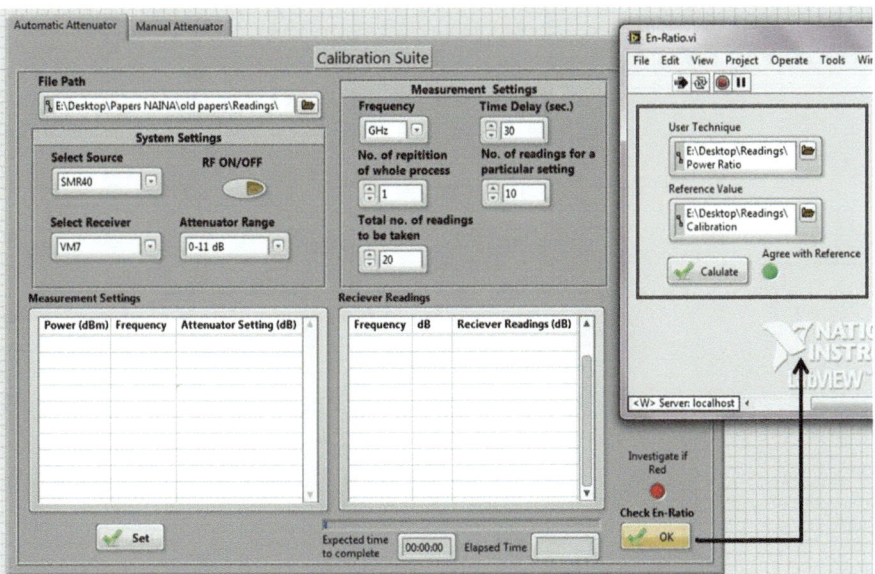

Fig. 5.3 Validation of results using E_n-ratio

5.3 Comparison of Results with Manual Measurements

The comparison of measured attenuation is shown in Fig. 5.4a–c. In attenuation measurement, 1, 5, and 10 dB attenuation steps have been assessed with an expanded uncertainty of ± 0.015 dB. It is a clear observation that for lower levels of attenuation and lower frequencies, a simple power ratio technique outperforms the function of a tuned receiver VM-7 and particularly at 60 MHz every technique

Fig. 5.4 Comparison of
Results

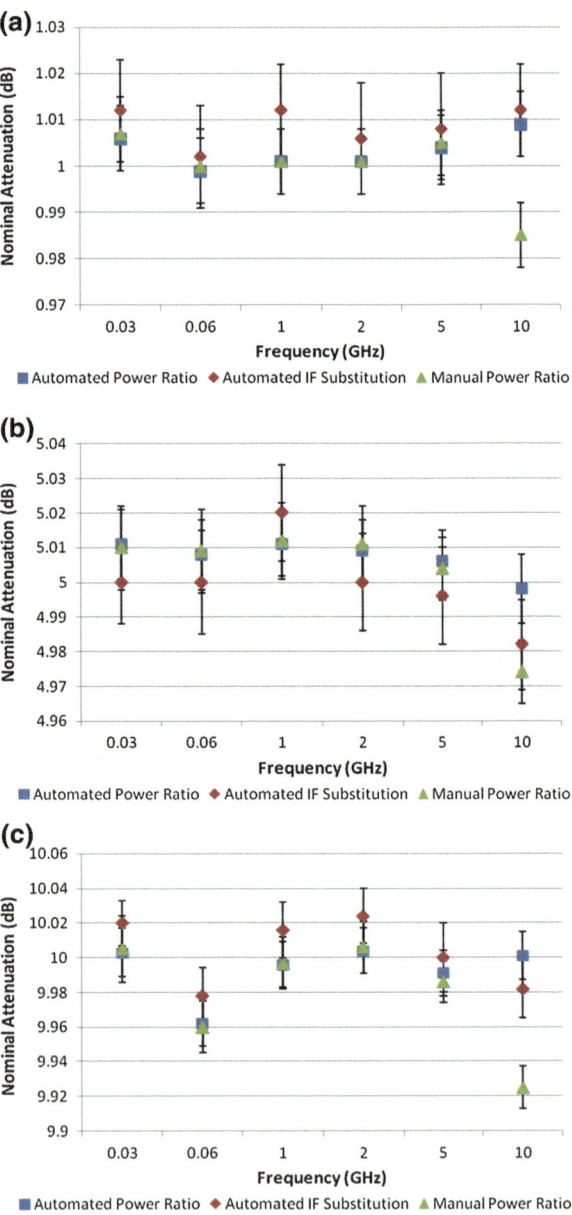

reads the same attenuation value. The results also show that 30 MHz IF substitution
technique gives an uncertainty of ±0.011 but power ratio technique is competent of
giving a better uncertainty of ±0.007 for lower level of attenuation. However, IF
substitution technique is used for higher level of attenuation measurement as a
tuned receiver has a higher dynamic range.

5.4   Report Generation

The report generation for metrological applications is a serious subject. The metrology laboratories that issue calibration and test reports to the customers should report all the details of the test or measurement conducted on the DUT. The following main considerations to be taken while expressing the measurement results:

- Description of the measurement process
- Major uncertainty contributors and their limits
- Statistical analysis of the results.

The previous chapters and sections have explained in detail the measurement process and uncertainty evaluation. Now if one has to generate the measurement report using LabVIEW, the NI Report Generation Toolkit can be used for generating the calibration report directly in a Microsoft Word or Excel file, as shown in Figs. 5.5 and 5.6. The figures show how the fundamental parameters of the calibration report can be pushed using LabVIEW for automated report generation. In a similar way, the measurement results acquired by the automated test setup can be filled up in the report as per the user format.

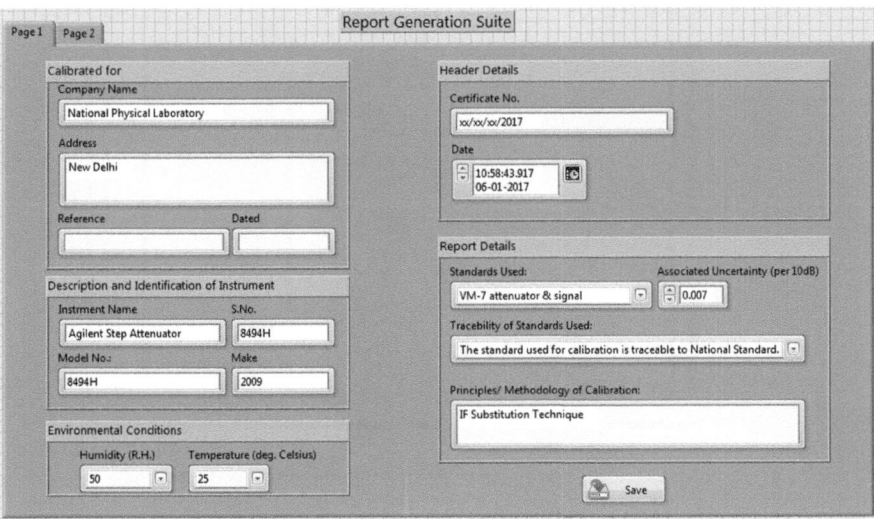

Fig. 5.5 Example report generation VI

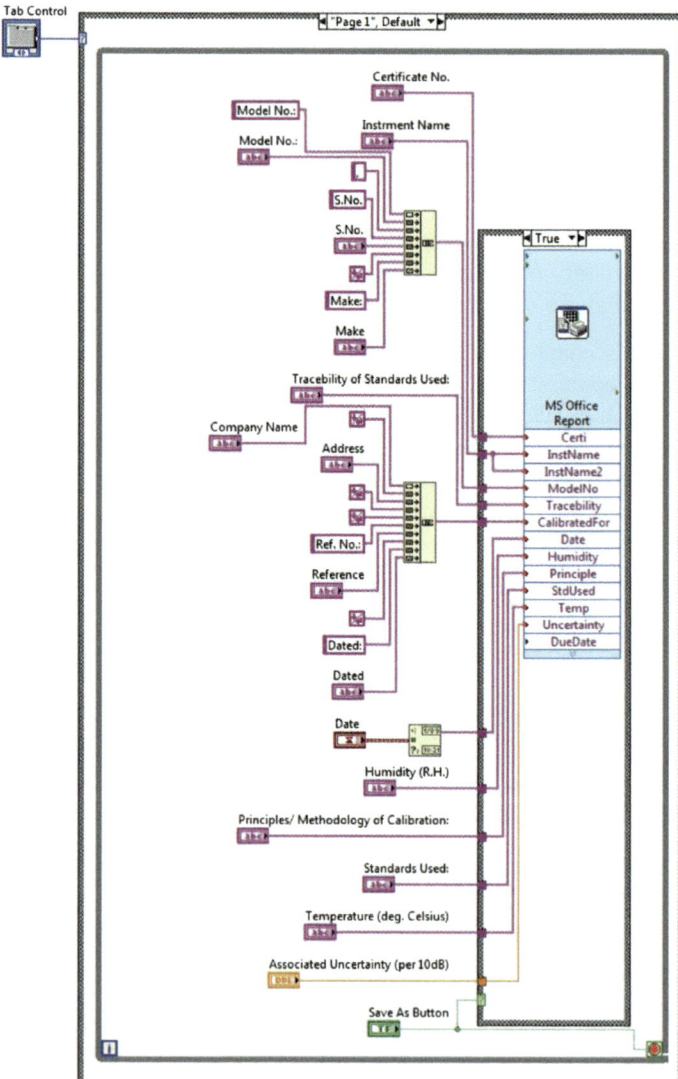

Fig. 5.6 Block diagram demonstrating report generation

Chapter 6
Conclusion

Here, LabVIEW programming for instrument control is employed to develop a software solution for metrological applications. Considering the common microwave measurements, examples are explained for understanding the LabVIEW programming techniques. The process for uncertainty evaluation is also given in detail to analyze the measured results. The observations made during the study and implementation can be summarized as follows:

(i) For instrument control, GPIB devices can be automated easily using LabVIEW message-based programming
(ii) Complete calibration suite for microwave measurements can possible provide faster tests with minimum human interference
(iii) Automated uncertainty evaluation can be implemented using LabVIEW Mathematics functions
(iv) The metrology software should in no case compromise with the integrity of the measurement results
(v) Automated report generation can be implemented using LabVIEW.

In a typical automated RF calibration process, the operator must frequently intervene to change test setups, thus limiting the benefits that can be realized by the automaton. However, the presented system of attenuation measurement is developed such that it only needs measurement points to be passed by the user just one time before the execution, thus increasing the calibration system capabilities in terms of time and efficiency. The measurement results along with the corresponding uncertainties show the proper implementation and execution of automation. The metrological software is validated with assured quality by following the guidelines available in the literature. It was finally found that the quality of the measured results is not affected by the automation since proper considerations were made during the design process. The LabVIEW application is also capable for measurement technique validation with the help of normalized error analysis. We hope this exercise will be useful for implementing automated attenuation measurement system in the measurement laboratories.

© The Author(s) 2018   43
S. Dubey et al., *LabVIEW Based Automation Guide for Microwave Measurements*,
SpringerBriefs in Computational Electromagnetics,
https://doi.org/10.1007/978-981-10-6280-3_6

About the Book

Modern metrology laboratory looks upon the automated test and measurement setups for better efficiency and throughput compared to manual procedures. NI LabVIEW can be used to develop the automated measurement systems using instrument control and data analysis. This book focuses on the automation of the microwave measurements. Before the software development for automation, common microwave measurements are discussed in detail. Some fundamentals about evaluation of uncertainty in these measurements are also discussed. Generic software is developed for calibration process of microwave parameters using message-based GPIB programming of instrument control. Example of attenuation measurement is demonstrated with highlight on the fundamental techniques required to understand LabVIEW programming. The reader would get an insight of software development for metrological applications. The book can be used as a guide for development of automation not just for attenuation but any of the other measurement process in readers' laboratory.

Uncited References

1. B. Wichmann, G. Parkin, R. Barker, *Software Support for Metrology Best Practice Guide No. 1, in Validation of Software in Measurement Systems* (National Physical Laboratory, Teddingtion, 2007)
2. P. Ciarlini, M.G. Cox, F. Pavese, D. Richter, *Advanced Mathematical and Computational Tools in Metrology VI*, Volume 66 of Series on Advances in Mathematics for Applied Sciences (World Scientific, Singapore, 2004). ISBN: 9814482412 2004
3. M.G. Cox, P.M. Harris, I.M. Smith, Software specifications for uncertainty evaluation. *National Physical Laboratory*
4. N. Greif, Software testing and preventive quality assurance for metrology. Comput. Stand. Interfaces **28**(3), 286–296 (2006)

© The Author(s) 2018
S. Dubey et al., *LabVIEW Based Automation Guide for Microwave Measurements*,
SpringerBriefs in Computational Electromagnetics,
https://doi.org/10.1007/978-981-10-6280-3

GPSR Compliance

*The European Union's (EU) General Product Safety Regulation (GPSR)
is a set of rules that requires consumer products to be safe and our
obligations to ensure this.*

*If you have any concerns about our products, you can contact us on
ProductSafety@springernature.com*

In case Publisher is established outside the EU, the EU authorized
representative is:

Springer Nature Customer Service Center GmbH
Europaplatz 3
69115 Heidelberg, Germany

Batch number: 09478958

Printed by Printforce, the Netherlands